ACTION MATH

SHAPES

Ivan Bulloch

Consultants
Wendy and David Clemson

WORLD BOOK

in association with

TWO-CAN

2 Shapes Around You

Everything you look at has a shape. Some things are curved. Some are straight. Some have points or corners. What words do you use to talk about the shapes around you?

Look at all the shapes on these pages. How many curved edges does each one have? How many straight edges does each one have?

Playing with Shapes

You can put some shapes side by side to make a completely different shape. You can also make patterns with shapes. Experiment with different shapes of building blocks.

The activities in this book will help you:
● spot and name shapes around you.
● learn some important facts about solid and flat shapes.
● make patterns using shapes.
● learn how to build and take apart shapes.

4 Printing Shapes

● Look for objects with interesting shapes that can be used to print a pattern or a picture. To print the pictures shown below, we used a sponge, a cork, an eraser, and a wooden building block. Find some other shapes to print. Get permission to use them.

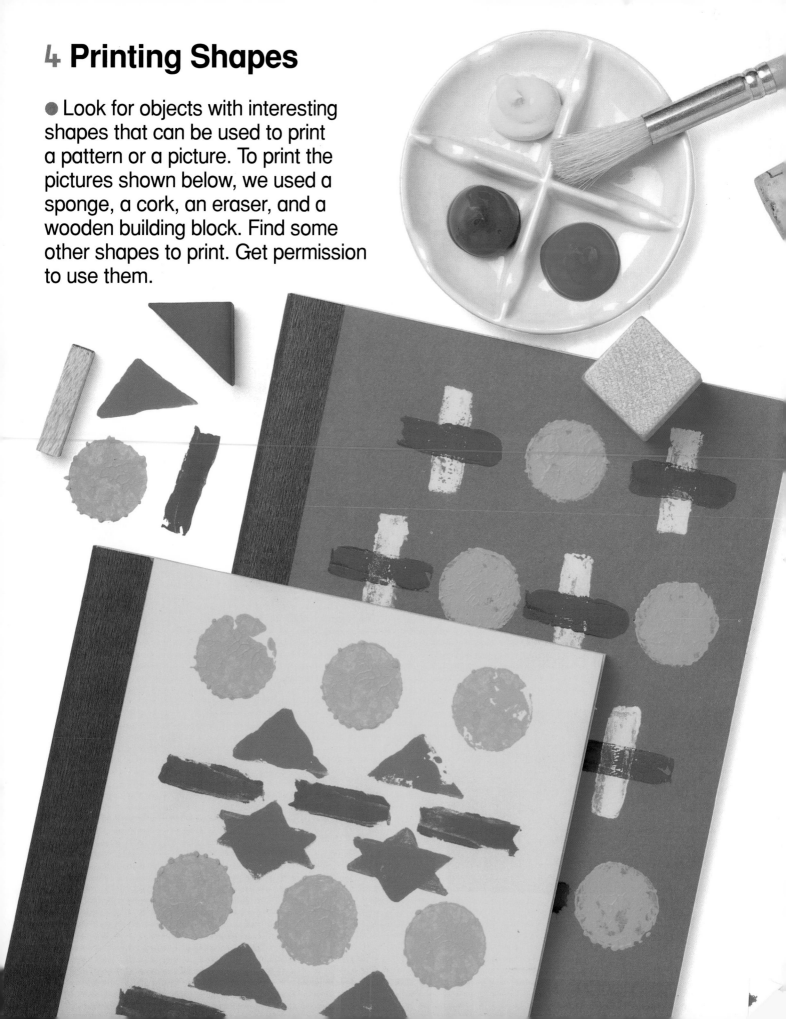

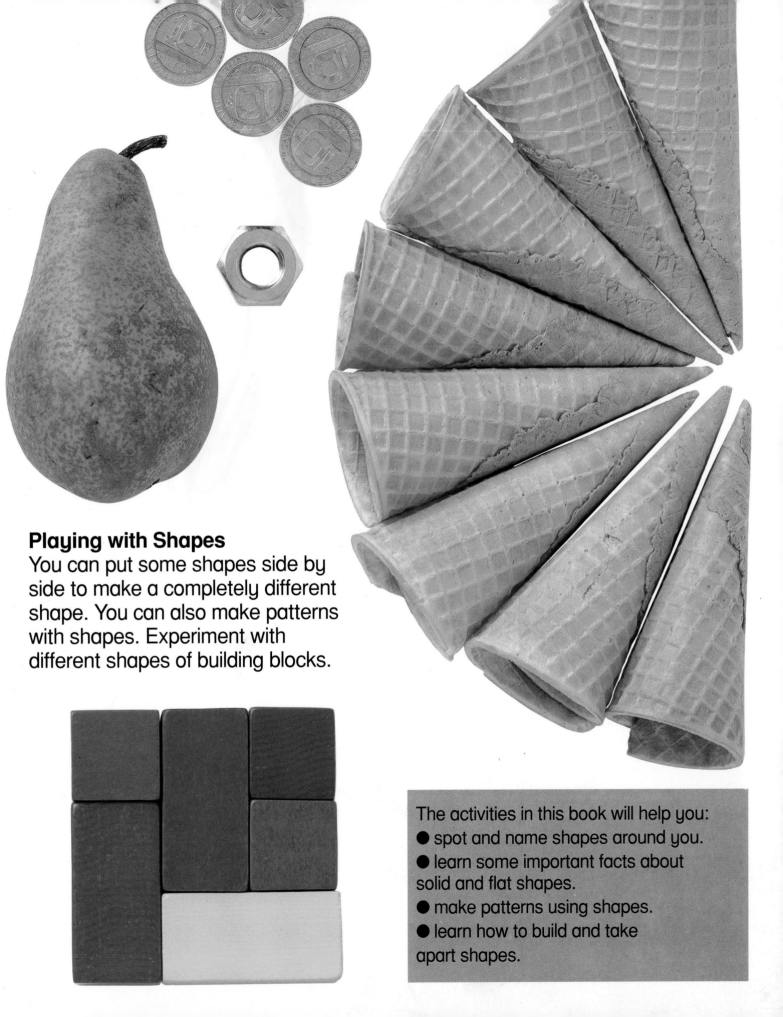

Playing with Shapes

You can put some shapes side by side to make a completely different shape. You can also make patterns with shapes. Experiment with different shapes of building blocks.

The activities in this book will help you:
- spot and name shapes around you.
- learn some important facts about solid and flat shapes.
- make patterns using shapes.
- learn how to build and take apart shapes.

4 Printing Shapes

● Look for objects with interesting shapes that can be used to print a pattern or a picture. To print the pictures shown below, we used a sponge, a cork, an eraser, and a wooden building block. Find some other shapes to print. Get permission to use them.

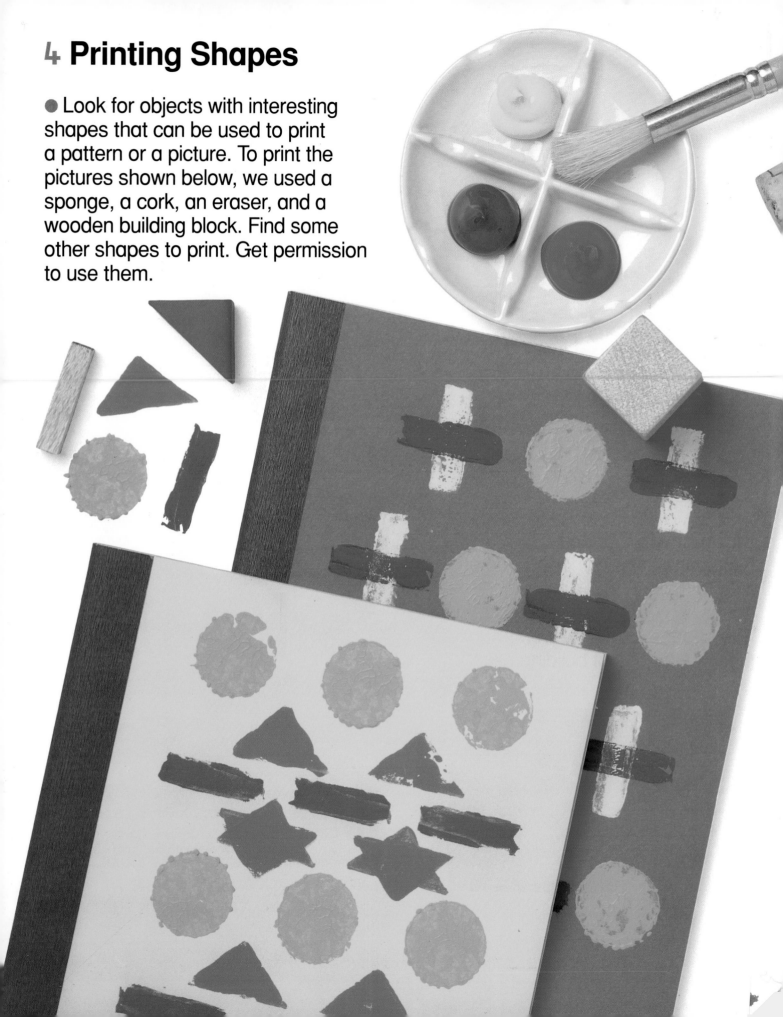

● Cover one side of your shape with paint. You could use a paintbrush or dip the shape into the paint. Press the shape carefully onto the paper.

● Some shapes can fit together with no gaps in between, like these squares. Which of your shapes fit together? Which shapes do not fit together?

Here's what you learn:
● how to recognize different shapes.
● how to fit shapes together.

6 Making Shapes

Here are some simple ways to make basic shapes from colored paper.

Square
● Start with a rectangle of paper.
● Place the paper so that one of the short ends is toward you.

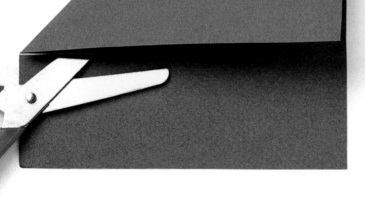

● Fold the top left corner toward you until the top of the paper lines up with the right-hand side of the paper.
● Cut off the single piece of paper left at the bottom.
● Unfold the paper and you will have a square.

Triangle
● Cut your square right down the diagonal fold to make two triangles.

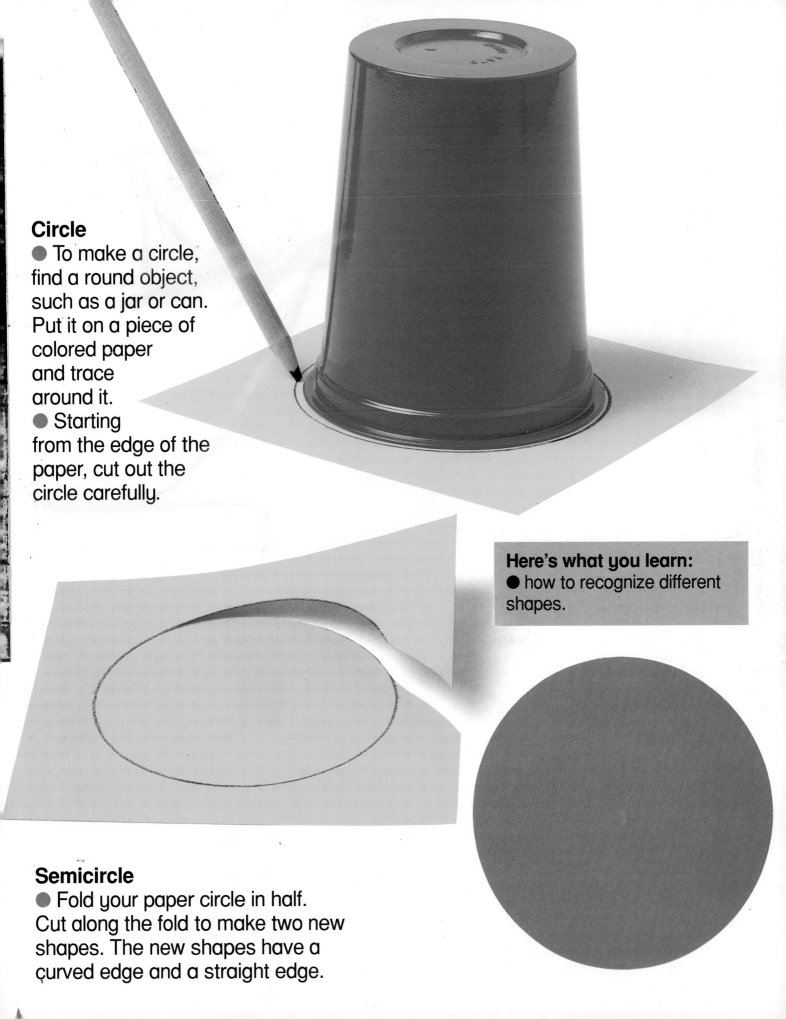

Circle

● To make a circle, find a round object, such as a jar or can. Put it on a piece of colored paper and trace around it.

● Starting from the edge of the paper, cut out the circle carefully.

Here's what you learn:
● how to recognize different shapes.

Semicircle

● Fold your paper circle in half. Cut along the fold to make two new shapes. The new shapes have a curved edge and a straight edge.

8 Shape Pictures

We made the picture here using colored paper shapes. You will be surprised at how easy it is to make an interesting picture of your own. Cut or tear many different paper shapes before you start.

Try It Out
● Arrange your shapes on a sheet of paper to make a picture.

● Don't glue the shapes down right away. Experiment with the shapes and move them around. When you are happy with your picture, carefully lift up each shape and glue it in place.

● Circles and ovals are good shapes for making heads and bodies. Triangles, rectangles, and squares are good for making buildings.

Here's what you learn:
● how to fit shapes together.
● how to use different shapes in design.

Patches

With just a few old scraps of fabric, you can make a colorful patchwork cloth.

● The six-sided shape below is called a hexagon. Trace it onto other paper and cut it out. Make lots of paper hexagons.

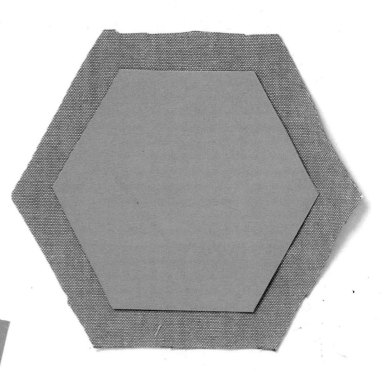

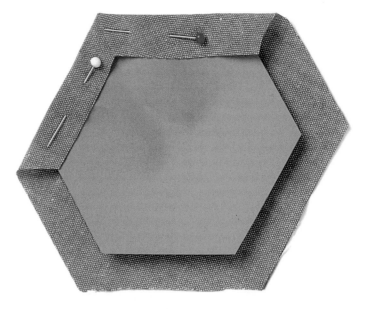

● Place each paper hexagon on a piece of fabric. Cut around the paper, leaving extra fabric on all sides.

● Fold each edge of the fabric and ask an adult to help you pin it to the edge of the paper. When you have pinned several pieces, you are ready to sew the hexagons together!

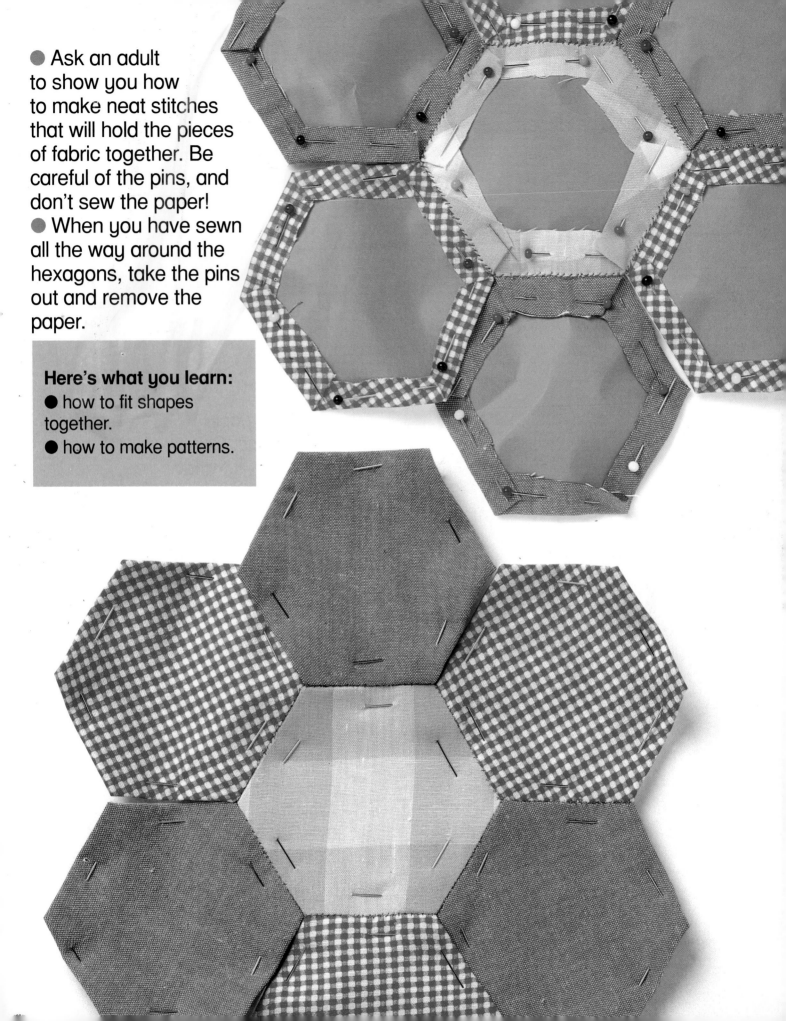

● Ask an adult to show you how to make neat stitches that will hold the pieces of fabric together. Be careful of the pins, and don't sew the paper!

● When you have sewn all the way around the hexagons, take the pins out and remove the paper.

Here's what you learn:
● how to fit shapes together.
● how to make patterns.

12 Cutout Shapes

You can make amazing shapes just by folding paper and making cutouts in the folded edge.

Fold and Cut

● Fold a piece of paper in half.

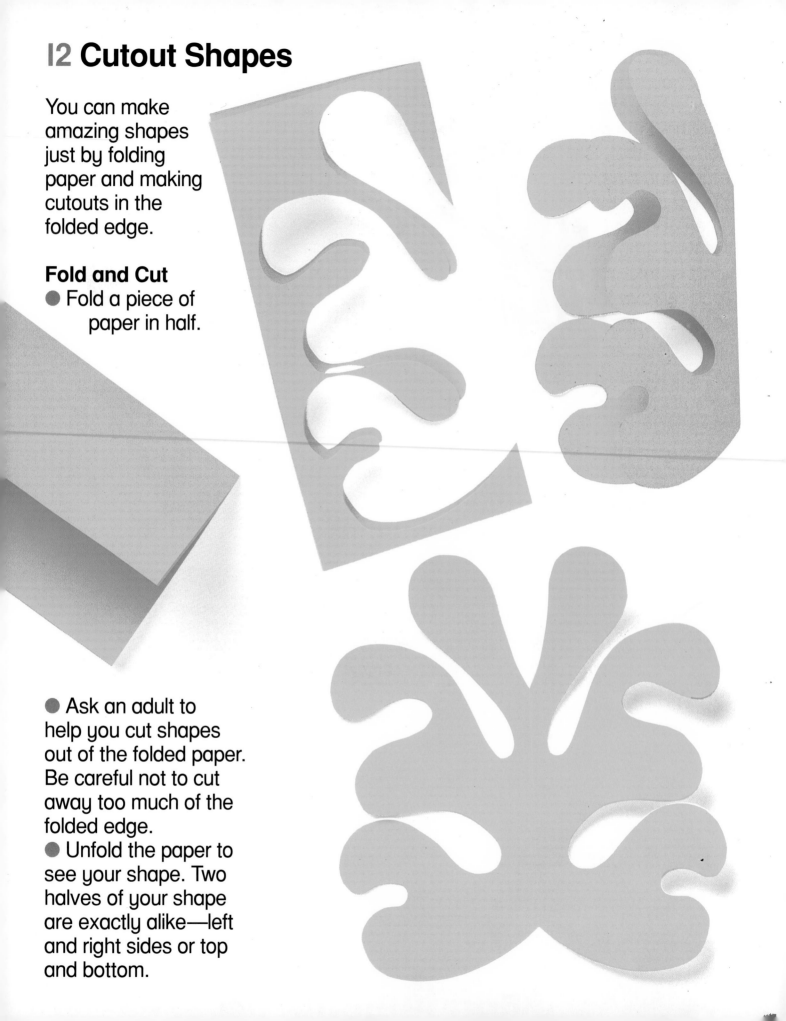

● Ask an adult to help you cut shapes out of the folded paper. Be careful not to cut away too much of the folded edge.

● Unfold the paper to see your shape. Two halves of your shape are exactly alike—left and right sides or top and bottom.

Four Folds

● Now fold another piece of paper in half, and then in half again.

● Cut shapes from both folded edges, but remember not to cut off too much!

● When you unfold the paper you will find that the right and left sides match, and the top and bottom match, too.

Here's what you learn:
● about symmetry.

14 Slit-and-Slot Shapes

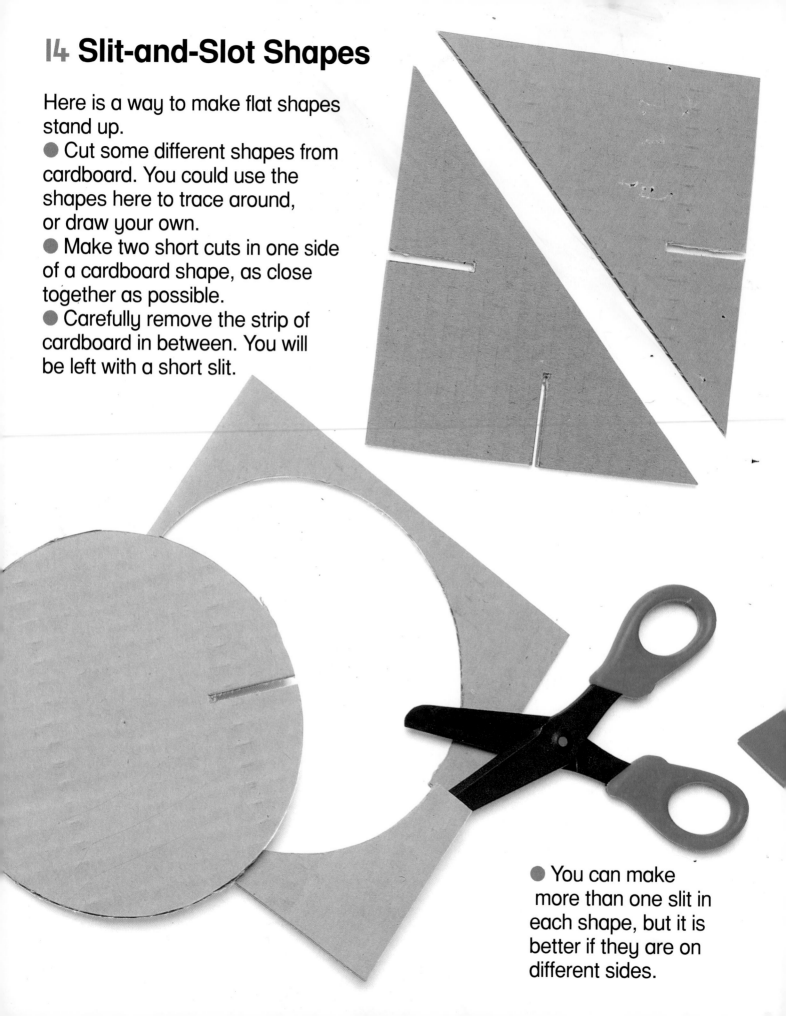

Here is a way to make flat shapes stand up.
● Cut some different shapes from cardboard. You could use the shapes here to trace around, or draw your own.
● Make two short cuts in one side of a cardboard shape, as close together as possible.
● Carefully remove the strip of cardboard in between. You will be left with a short slit.

● You can make more than one slit in each shape, but it is better if they are on different sides.

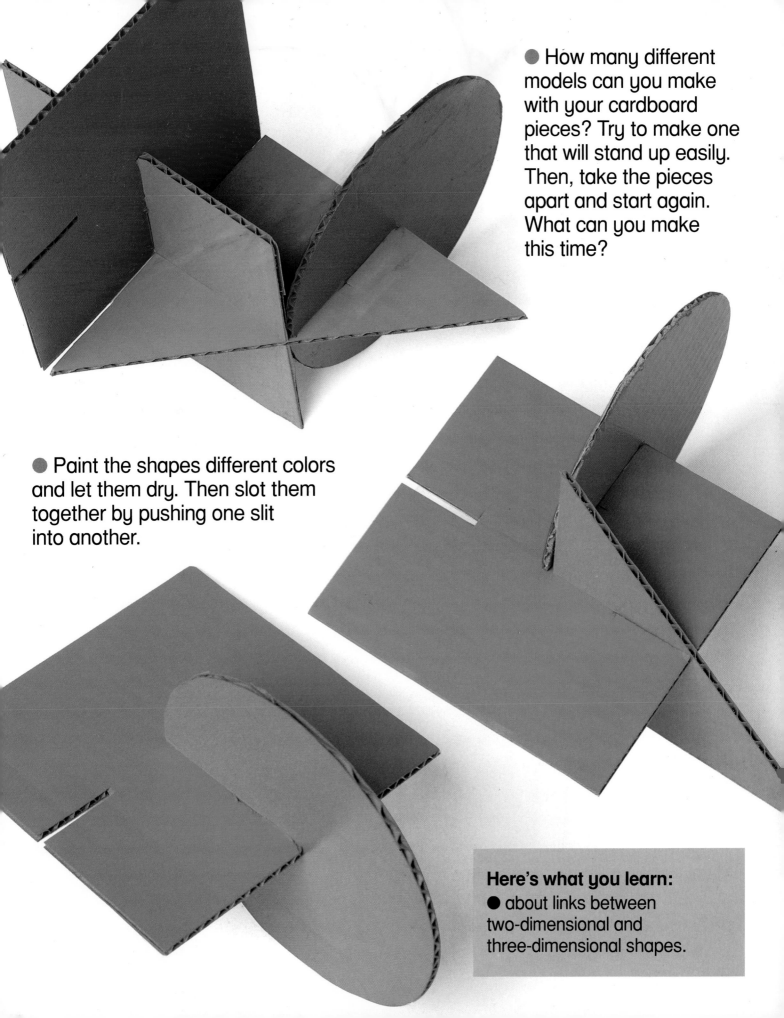

● How many different models can you make with your cardboard pieces? Try to make one that will stand up easily. Then, take the pieces apart and start again. What can you make this time?

● Paint the shapes different colors and let them dry. Then slot them together by pushing one slit into another.

Here's what you learn:
● about links between two-dimensional and three-dimensional shapes.

Once you know how to slit and slot, you can make all sorts of models that stand up by themselves. Use thick paper or thin cardboard to make your models.

Slit-and-Slot Bushes

● Draw two bush shapes roughly the same size and cut them out. Cut a slit from the bottom of one shape to the middle. Then, cut a slit from the top of the other shape to the middle. Paint the shapes bright colors. Slot them together and stand the bush up.

Tall Trees

● Draw two tree shapes. You could use the tree here as a guide to trace around. Cut the shapes out. Make slits in both trees, as you did for the bushes. Now stand the tree up.

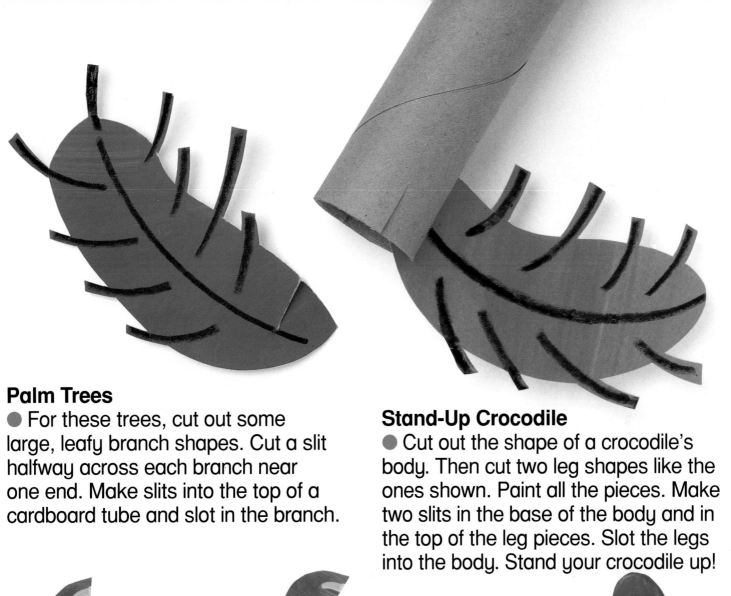

Palm Trees

● For these trees, cut out some large, leafy branch shapes. Cut a slit halfway across each branch near one end. Make slits into the top of a cardboard tube and slot in the branch.

Stand-Up Crocodile

● Cut out the shape of a crocodile's body. Then cut two leg shapes like the ones shown. Paint all the pieces. Make two slits in the base of the body and in the top of the leg pieces. Slot the legs into the body. Stand your crocodile up!

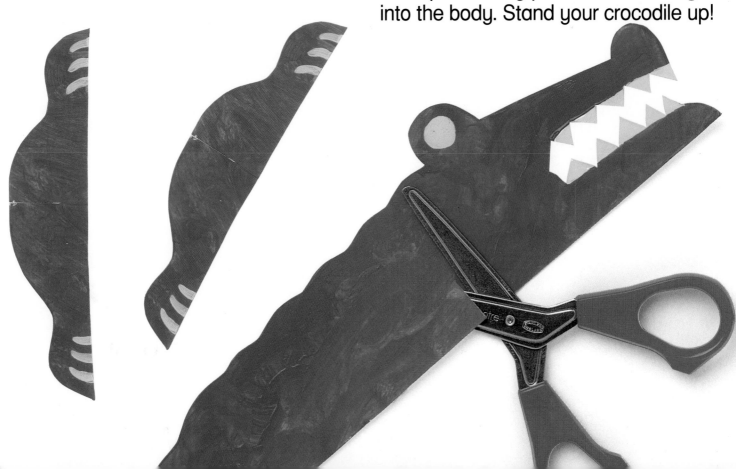

Now arrange your trees and bushes to make a jungle scene. You can make tall or short trees by using different sized cardboard tubes.

Hanging Bird
● Make a bird to keep the crocodile company. You will need three pieces — one for the body, one for the wings, and one for the head.

● Try making other kinds of plants. Can you think of any other slit-and-slot animals to put in your jungle?

20 Food Monster

Look at all the different shapes of vegetables on this page. Some of them are very strange! Can you think of a way to describe them? Try making a funny food monster to decorate your table. Use round shapes to make the eyes and nose, and long shapes for the arms and legs.

● Ask a grown-up to help you join the vegetables together with toothpicks.
● Push one end of a toothpick into a vegetable.
● Then, push another vegetable onto the other end of the toothpick.

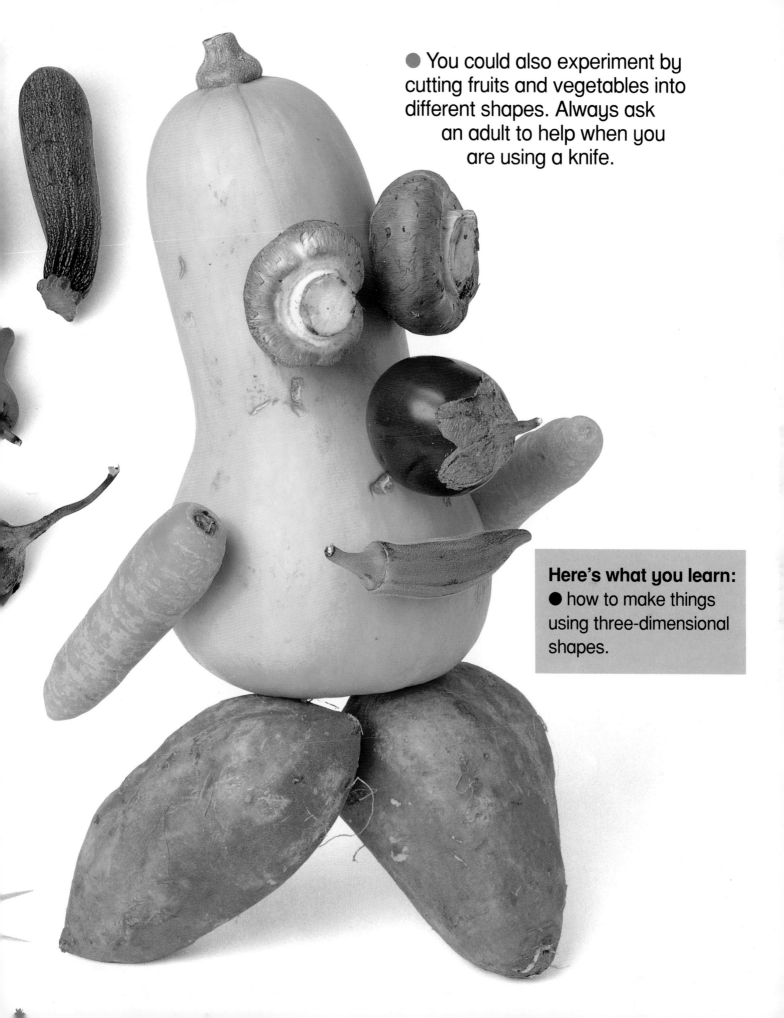

● You could also experiment by cutting fruits and vegetables into different shapes. Always ask an adult to help when you are using a knife.

Here's what you learn:
● how to make things using three-dimensional shapes.

22 Looking at 3-D Shapes

Containers come in lots of shapes and sizes. The next time you are in a store, see how many different shapes you can spot.

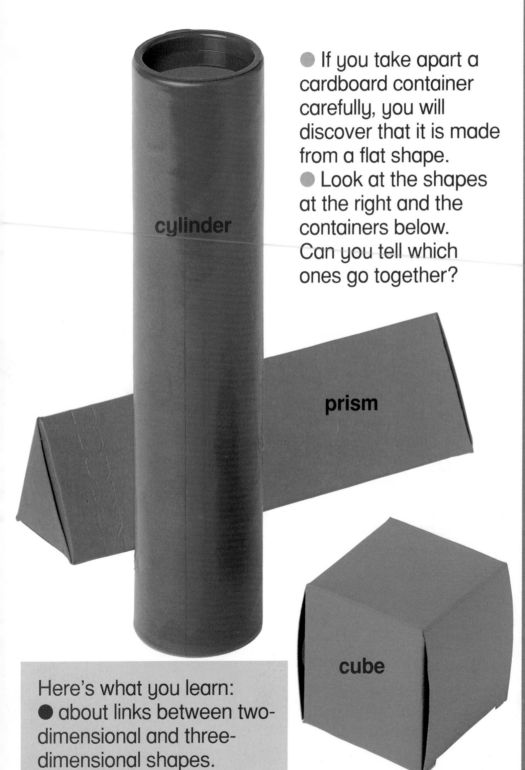

cylinder

prism

cube

● If you take apart a cardboard container carefully, you will discover that it is made from a flat shape.
● Look at the shapes at the right and the containers below. Can you tell which ones go together?

Here's what you learn:
● about links between two-dimensional and three-dimensional shapes.

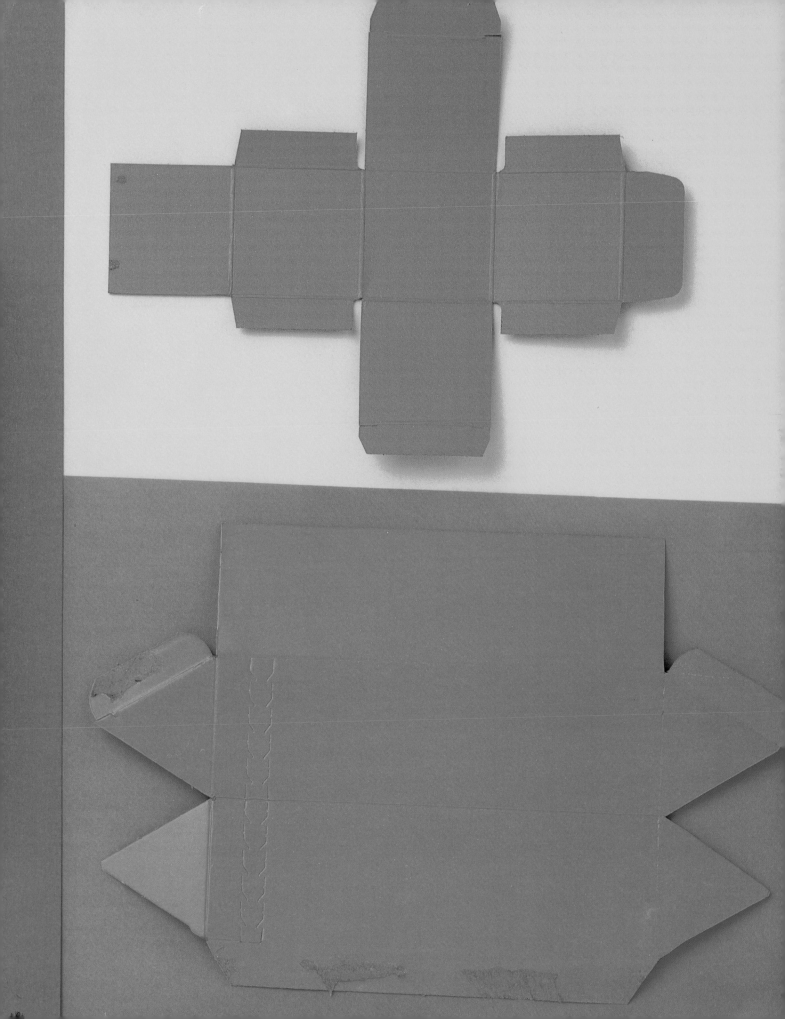

24 Make a Cube

On pages 22 and 23 are some containers that are opened out. If you follow the instructions here, you will be able to turn a flat piece of cardboard into a cube.

● Trace the shape on this page. Use your tracing to cut out the same shape from a piece of thin cardboard.

● Use a pencil and ruler to draw in the dotted lines, using the picture as a guide.

● You need to make the shape as precisely as possible so that all the sides of your cube will fit together.

The three narrow strips with shaped corners are called tabs.

● Fold the cardboard along each of the dotted lines, keeping the pencil marks on the inside of the cube.

● Fold the longest strip over to meet the tab on the opposite side.
Glue the tab and press it firmly onto the strip from the inside. Then, glue one of the tabs that is on a shorter side. Tuck the tab in and press it from the inside. Fold the last flap up to make the lid of the box.
Tuck it in without gluing.

Here's what you learn:
● about the flat shapes that make up a container.
● how to draw and cut precisely.

26 Desk Organizer

You should be able to find cardboard tubes in lots of different shapes and sizes. They are often used for packaging, or inside rolls of paper and aluminum foil. Collect some tubes and turn them into a useful storage unit for your pencils, scissors, and erasers.

● Arrange your painted tubes in a group and glue them together along the sides.

Sorting
● First, decide which tubes are best for the things you want to store in them. For example, your pencils may need a long, thin tube. If necessary, you could ask an adult to cut a tube to make it shorter.
● Paint the tubes bright colors.

● Make a base for your storage unit by gluing the bases of the tubes to a piece of cardboard. Fill the tubes with all your pencils, pens, and crayons.

Here's what you learn:
● how to recognize shapes in different sizes.

28 Build an Animal

Our colorful animal was made from used boxes, a ball, and tubes. Start collecting as many shapes as you can. Get permission to use the things you collect. Look for interesting shapes with curved and straight edges. If you are using containers that have had food in them, make sure you clean them first.

● You can use sections cut from boxes to make different shapes. We made the triangle out of the corner of a box and the round shape (a sphere) is a painted foam ball.

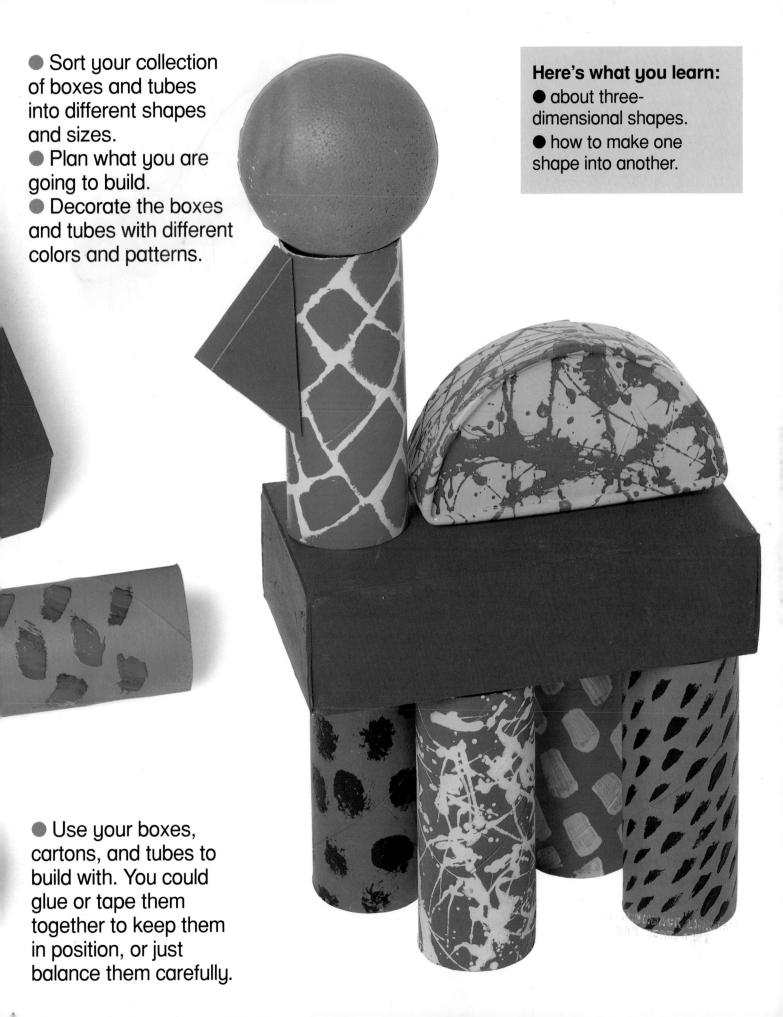

● Sort your collection of boxes and tubes into different shapes and sizes.

● Plan what you are going to build.

● Decorate the boxes and tubes with different colors and patterns.

● Use your boxes, cartons, and tubes to build with. You could glue or tape them together to keep them in position, or just balance them carefully.

30 Stretchy Jewelry

Here is a way to turn flat pieces of paper into completely different shapes that you can wear. Your friends will be amazed! You will need long strips of paper in two different colors.

Accordion Folds

● Glue the ends of two strips of paper together at right angles. Look at the picture at the top.

● Then, fold the strip that is underneath over the top strip.

● Next, fold the strip which is now underneath over the top one.

● Continue folding until you get to the ends of the strips.

● Glue the ends down.

● Gently pull the two ends of the paper to open the chain a little. When you let go, the shape will spring back. To make a longer piece, glue extra strips of paper onto the end of the chain.

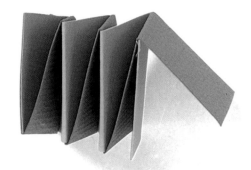

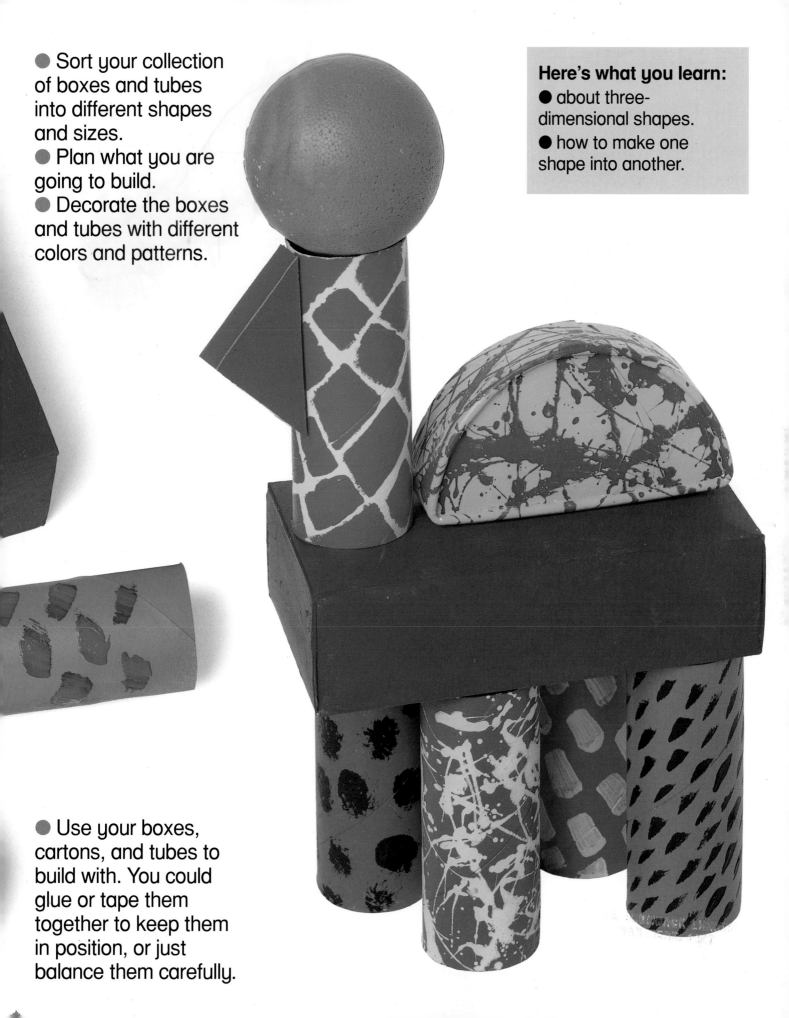

● Sort your collection of boxes and tubes into different shapes and sizes.

● Plan what you are going to build.

● Decorate the boxes and tubes with different colors and patterns.

● Use your boxes, cartons, and tubes to build with. You could glue or tape them together to keep them in position, or just balance them carefully.

30 Stretchy Jewelry

Here is a way to turn flat pieces of paper into completely different shapes that you can wear. Your friends will be amazed! You will need long strips of paper in two different colors.

Accordion Folds
● Glue the ends of two strips of paper together at right angles. Look at the picture at the top.
● Then, fold the strip that is underneath over the top strip.
● Next, fold the strip which is now underneath over the top one.
● Continue folding until you get to the ends of the strips.
● Glue the ends down.

● Gently pull the two ends of the paper to open the chain a little. When you let go, the shape will spring back. To make a longer piece, glue extra strips of paper onto the end of the chain.

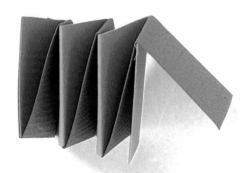

You can use your stretchy paper chains to make colorful necklaces, earrings, and bracelets.

Here's what you learn:
● how to turn a two-dimensional shape into a three-dimensional one.
● how to make right angles.

Editor: Diane James
Design: Beth Aves
Text: Claire Watts
Photography: Toby
Editorial Assistant: Jacqueline McCann

Published in the United States and Canada by
World Book, Inc.
525 W. Monroe Street
Chicago, IL
60661
in association with Two-Can Publishing Ltd.

**For information on other World Book products,
call 1-800-255-1750, x 2238,
or visit us at our Web site at http://www.worldbook.com**

Library of Congress Cataloging-in-Publication Data

Bulloch, Ivan.
    Shapes / Ivan Bulloch; consultants, Wendy and David Clemson.
        p.   cm. – (Action math)
    Originally published: New York: Thomson Learning, 1994.
    Includes index.
    Summary: Teaches the principles of symmetry, tesselation, and two-dimensional and three-dimensional shapes by means of various handicrafts.
    ISBN 0-7166-4904-7 (hardcover)—ISBN 0-7166-4905-5 (softcover)
    I. Geometry–Juvenile literature.   [I. Geometry.   2. Handicraft.
3. Shape.]   I. Clemson, Wendy.   II. Clemson, David.   III. Title.
IV. Series: Bulloch, Ivan.   Action math.
QA445.5.B86   1997
516'.15–dc21   96-49558

Printed in Hong Kong

2 3 4 5 6 7 8 9 10 01 00 99 98 97

Skills Index

Consultants
Wendy and David Clemson are experienced teachers and researchers. They have written many successful books on mathematics.